Fracktal Mathematics:

A Paradigm Shift in Infinity, Complexity, and Interdisciplinary Synthesis

I0422231

By: Gregory J. Betti

Abstract

Fracktal Mathematics represents a paradigm shift in mathematical thought, challenging conventional

notions of infinity and complexity. This thesis is a comprehensive exploration of Fracktal Mathematics,

examining its core principles, interdisciplinary applications, and trans-formative potential. By juxtaposing

traditional mathematical equations with their Fracktal counterparts, we illustrate the power of this new

approach. From physics to technology, we venture into the uncharted territories where Fracktal

Mathematics redefines relationships, interconnects systems, and invites a symphony of possibilities.

CHAPTER 1:
INTRODUCTION

In this chapter, we set the stage for our exploration by discussing the motivations, objectives, and scope

of this thesis. We examine the limitations of traditional mathematics in addressing complex systems and

introduce the revolutionary concept of Fracktal Mathematics1.1 Background and Motivation

The evolution of human thought has been marked by a relentless pursuit of understanding the

intricacies of the universe. Mathematics, as the language of this exploration, has provided a framework

to decode and decipher the hidden patterns underlying the natural world. However, the limits of

traditional mathematical paradigms in addressing the complexity of interconnected systems have spurred the birth of a new era of inquiry — Fracktal Mathematics.

In an age where technology has enabled us to glimpse the cosmos in unprecedented detail and

complexity, conventional mathematical models have often fallen short in capturing the essence of

emergent behavior. The quest for a more nuanced, holistic, and dynamic approach has led us to the

threshold of Fracktal Mathematics. The motivation behind this exploration lies in the pursuit of a

mathematical language that mirrors the interconnected nature of the universe, inviting us to rethink

infinity, complexity, and the very foundations of mathematics.

1.2 Research Objectives

This thesis embarks on a journey to unravel the core principles of Fracktal Mathematics and to illustrate

its potential across a diverse spectrum of scientific disciplines and technological landscapes. The

objectives of this research are manifold:

To comprehend the essence of Fracktal Mathematics and its departure from conventional mathematical

constructs.

To showcase the transformative impact of Fracktal Mathematics on traditional equations and established

theories, offering a visual and intuitive understanding of the paradigm shift.

To explore how Fracktal Mathematics redefines the boundaries of interdisciplinary exploration, from

physics to economics, and from technology to philosophy.

To shed light on the implications of Fracktal Mathematics for education, collaborative research, and

ethical considerations.

To envision future directions in research and application, envisioning a world where Fracktal

Mathematics serves as a catalyst for innovation and holistic understanding.

1.3 Scope and Limitations

While this study delves into the foundational principles and potential applications of Fracktal

Mathematics, it is important to acknowledge certain limitations. The practical implementation of Fracktal

Mathematics in various fields may require specialized tools and methodologies that are yet to be fully

developed. Additionally, the philosophical and ethical implications of this new mathematical paradigm

are subjects that merit in-depth exploration beyond the scope of this thesis. As we embark on this intellectual journey, we invite the reader to consider the possibility of a

mathematical world that transcends boundaries, embraces complexity, and redefines the very essence

of interconnectedness. In the following chapters, we will journey through the uncharted territories of

Fracktal Mathematics, unveiling its beauty, impact, and potential to revolutionize our understanding of

the cosmos.

CHAPTER 2:
REDEFINING INFINITY:
THE CORE TENET
OF FRACKTAL
MATHEMATICS

Here, we delve deep into the heart of Fracktal Mathematics — its redefinition of infinity. We explore the

meta-equation "Infinity — Infinity = x" and the implications of considering infinity as a dynamic,

interactive entity. Fracktal Mathematics re-imagines the canvas of mathematical exploration, where

complexity is not a puzzle to be solved but an artistic tapestry of emergent patterns.

2.1 Infinity — Infinity = x: A Dynamic Equation

At the core of Fracktal Mathematics lies an equation that

challenges our conventional understanding of infinity: "Infinity — Infinity = x." This equation stands as a testament to the dynamic nature of infinity itself, redefining it from an abstract concept to a force that interacts and transforms within the mathematical realm.

In traditional mathematics, infinity has often been treated as a static, unchanging concept — a value that is unreachable and beyond calculation. However, Fracktal Mathematics introduces a paradigm shift by asserting that the difference between two infinities can manifest as any conceivable traditional equation. This seemingly paradoxical notion disrupts our traditional view of infinity, inviting us to explore its fluid and interactive nature.

Consider, for instance, the simple equation $8 \times 8 = 64$. In the realm of Fracktal Mathematics, this equation takes on a new dimension of meaning. It evolves into an exploration of infinity's creative potential:

$$1 \div (1/\text{Infinity}) = 8 \times 8 = 64.$$

This transformation demonstrates how the symphony of interconnectedness, which defines Fracktal Mathematics, can reframe even the most basic mathematical relationships. In this new interpretation, conventional equations become windows into the infinite variations that arise from the interplay of infinity with other mathematical constructs.

The equation "Infinity — Infinity = x" challenges us to abandon our preconceived notions and engage in a deeper dialogue with mathematical reality. It beckons us to question the rigid boundaries we have imposed on infinity and encourages us to embrace its dynamic, ever-changing character. Through this equation, Fracktal Mathematics empowers us to explore mathematical relationships as a canvas for creative expression, where infinity's infinite potential takes center stage.

As we traverse the terrain of Fracktal Mathematics, we begin to appreciate that even the most fundamental mathematical elements are not set in stone, but rather partake in a dynamic dance that mirrors the intricate interconnectedness of the universe. In the following sections of this chapter, we will delve further into the implications of this dynamic equation and explore how it re-frames our understanding of mathematics, infinity, and complexity.

CHAPTER 2.2: EMBRACE OF COMPLEXITY'S INTRICACIES

Fracktal Mathematics is not solely defined by its re-imagining of infinity; it is equally distinguished by its embrace of complexity as an inherent feature of the universe. Unlike traditional mathematical models, which often strive to simplify and reduce complex phenomena to linear relationships, Fracktal Mathematics boldly recognizes complexity as a symphony of interconnected patterns that emerge through non-linear scaling.

Conventional mathematical approaches often adhere to linear methods, seeking to break down intricate systems into manageable components and linear equations. However, Fracktal Mathematics challenges this reductionist approach by celebrating the inherent interconnectedness of systems, where each component resonates and interacts with others in intricate ways. This perspective offers a richer, more nuanced understanding of how complex phenomena unfold.

Imagine a fractal, that remarkable geometric pattern known for its self-replicating properties across various scales. Much like the fractal's intricate details that remain consistent regardless of magnification, Fracktal Mathematics explores systems where complexity manifests at various levels. This approach allows us to unveil emergent properties that remain concealed in traditional mathematical frameworks.

In embracing complexity's intricacies, Fracktal Mathematics illuminates the hidden symmetries and patterns that shape our world. It encourages us to view intricate systems not as puzzles to be solved through reduction, but as dynamic tapestries woven from the threads of interconnectedness. Through this lens, complex phenomena, whether observed in physics, biology, economics, or any other field, become vibrant expressions of the underlying interconnected symphony.

As we journey through the landscapes of Fracktal Mathematics, we are invited to embrace the richness of complexity and to marvel at the harmonious, interconnected dance of emergent patterns. The equations that emerge in this paradigm offer more than solutions; they present us with an opportunity to explore the universe's hidden symmetries and to celebrate the beauty that arises from its intricate interconnectedness.

CHAPTER 3: FRACKTAL MATHEMATICS IN DIVERSE SCIENTIFIC DISCIPLINES

This chapter unfolds the versatility of Fracktal Mathematics across a spectrum of scientific disciplines, showcasing its profound impact on various domains. Through visual representation and comparisons between conventional and Fracktal equations, we unveil the trans-formative potential of this mathematical paradigm within the realms of physics, quantum mechanics, biology, and economics.

CHAPTER 3.1: PHYSICS: QUANTUM DYNAMICS BEYOND THE LINEAR

The marriage of Fracktal Mathematics with the intricate world of physics unveils a paradigm-shifting perspective that transcends traditional linear equations. Within this newfound approach, the intricate dance of quantum particles takes on a profound new meaning, echoing the symphony of interconnectedness that underpins the fabric of the universe.

Quantum Mechanics and the Fracktal Framework

The realm of quantum mechanics, with its entangled particles and probabilistic behavior, finds an ideal companion in Fracktal Mathematics. Traditional physics equations often deal with linear relationships and deterministic outcomes, yet quantum mechanics challenges these conventions with its inherent uncertainty and non-local correlations. Fracktal Mathematics becomes a lens through which we can explore the intricate interplay of quantum states, offering a more holistic and intuitive perspective on these phenomena.

The Schrödinger Equation Re-imagined

At the heart of quantum mechanics lies the Schrödinger equation, a fundamental equation that describes the behavior of particles at the quantum level. In the context of Fracktal Mathematics, this equation evolves beyond its linear boundaries. Rather than depicting particles' behavior in isolation, the equation becomes a canvas for exploring the interconnected symphony of quantum states.

Consider an electron's orbital behavior around an atomic nucleus. Traditional linear equations describe discrete energy levels and orbits. However, in the realm of Fracktal Mathematics, these energy levels become interconnected nodes, creating an intricate web of quantum possibilities. This interconnectedness reflects the true nature of particles, where their behavior is deeply influenced by the presence of other particles and their states.

Quantum Entanglement and Emergent Patterns

One of the most baffling phenomena in quantum mechanics is entanglement, where particles become inseparably linked regardless of distance. In the Fracktal framework, entanglement transforms from a mysterious phenomenon to an exploration of interconnected probabilities.

Imagine two entangled particles with opposite spins. Conventional physics equations describe this phenomenon with linear correlations. However, Fracktal Mathematics dives deeper, illustrating how the spins of these particles are part of a larger symphony — a symphony of interconnected possibilities that unfold when measured. This interconnectedness hints at the hidden harmony underlying seemingly paradoxical phenomena.

Exploring Quantum Waves and Beyond

Fracktal Mathematics also invites us to explore the wave-particle duality — the intriguing notion that particles exhibit both wave-like and particle-like behavior. In traditional equations, this duality is often treated as a binary concept. However, Fracktal Mathematics visualizes this duality as a spectrum of interconnected behaviors, where wave functions interact and merge to create emergent patterns of behavior.

Consider the famous double-slit experiment, where particles exhibit wave-like interference patterns. Fracktal Mathematics transforms this experiment into an exploration of the intertwined nature of wave functions, as they interact and create a tapestry of probabilities. This interconnected symphony of waves, akin to the resonance of musical notes, offers a deeper understanding of the dual nature of particles.

As we journey through the realm of quantum mechanics with the aid of Fracktal Mathematics, we witness the emergence of a more intuitive and interconnected interpretation. This perspective does not negate the precision of traditional quantum equations; instead, it enriches our understanding by revealing the symphonic tapestry that underlies the behavior of particles.

CHAPTER 3.2: QUANTUM MECHANICS: INTERPLAY OF QUANTUM STATES IN FRACKTAL FRAMEWORK

Within the enigmatic realm of quantum mechanics, where uncertainty and non-locality prevail, Fracktal Mathematics offers an innovative lens through which to understand the interconnected nature of quantum phenomena. This chapter delves into the profound transformation that occurs when traditional linear equations give way to the emergent symphonies of Fracktal Mathematics within the domain of quantum mechanics.

A Quantum Dance of Interconnectedness

Quantum mechanics has long perplexed scientists with its

behavior-defying properties. Fracktal Mathematics, with its emphasis on interconnectedness and emergent patterns, provides a fresh perspective on the underlying symphony of quantum behavior. Traditional linear equations often treat particles as isolated entities, while Fracktal Mathematics guides us toward an understanding of how particles interact as interconnected actors on a cosmic stage.

Embracing Quantum Entanglement

At the heart of quantum interconnectedness lies the phenomenon of entanglement. In traditional physics equations, entanglement can appear as a puzzling correlation between particles with no apparent communication. However, within the Fracktal framework, entanglement becomes a natural consequence of interconnectedness.

Imagine two entangled particles separated by vast distances. In Fracktal Mathematics, these particles are not isolated entities but rather nodes within a vast network of probabilities. When one particle's state is measured, the interconnected symphony resonates instantaneously across space, influencing the state of its entangled partner. Fracktal Mathematics allows us to perceive entanglement not as an anomaly but as a testament to the interconnected dance of particles.

Interwoven Probabilities and Wave Functions

Fracktal Mathematics also sheds light on the elusive wave-particle duality — a central feature of quantum behavior. Instead of viewing particles' behaviors as either waves or particles, Fracktal Mathematics invites us to explore the spectrum of interconnected behaviors that emerge from the interplay of quantum states.

Consider the Young's double-slit experiment, where particles exhibit interference patterns like waves. In the Fracktal framework, this experiment becomes a visual exploration of the intricate interweaving of wave functions. The emergent patterns that arise as these waves interact embody the interconnected nature of quantum behavior, offering a deeper understanding of how particles traverse this spectrum between particle-like and wave-like behaviors.

Uncertainty and the Fractured Symphony

Uncertainty, a fundamental principle of quantum mechanics, finds resonance in the Fracktal framework. Instead of treating uncertainty as a limitation, Fracktal Mathematics portrays it as an invitation to explore the myriad interconnected possibilities inherent in quantum systems.

Imagine the Heisenberg uncertainty principle, which limits our precision in measuring certain pairs of complementary properties. Fracktal Mathematics transforms this limitation into a canvas of interconnected probabilities. Instead of searching for precise values, we embark on a journey through the symphony of possibilities, where uncertainty becomes a feature rather than a hindrance.

As we traverse the landscapes of quantum mechanics through the lens of Fracktal Mathematics, the universe's hidden melodies come alive. This perspective not only enriches our understanding of quantum behavior but also challenges us to embrace the interconnected tapestry that shapes the quantum realm.

CHAPTER 3.3: BIOLOGY: ECOSYSTEMS UNVEILED THROUGH FRACKTAL PATTERNS

Biology, a tapestry of life interwoven with complexity, finds resonance in the realm of Fracktal Mathematics. This chapter delves into how the intricate relationships within ecosystems are illuminated when viewed through the lens of interconnectedness and emergent behavior. Fracktal Mathematics becomes a symphony of patterns, inviting us to explore the harmonious and chaotic rhythms that define the biological world.

Ecosystems as Dynamic Networks

In the realm of biology, ecosystems represent a delicate balance of interconnected relationships between species, environments, and resources. Traditional mathematical models often simplify these relationships, leading to an incomplete understanding of the intricate dynamics that govern life on Earth. Fracktal Mathematics revolutionizes this perspective by capturing the true essence of ecosystems as dynamic and interconnected networks.

Imagine a predator-prey relationship within an ecosystem.

Traditional linear equations may focus solely on the population dynamics of predator and prey species. However, within the Fracktal framework, these equations give way to a symphony of interconnected nodes representing various species, each influencing the others' population dynamics. The result is a dance of emergent patterns — a harmonious ballet that arises from the interplay of myriad species and their complex interactions.

Interwoven Adaptation and Emergence

Fracktal Mathematics invites us to explore how emergent properties arise from the interconnected web of life. Within ecosystems, species adapt to changing conditions, leading to emergent behaviors that transcend simple cause-and-effect relationships. Fracktal Mathematics allows us to view adaptation not as isolated events but as part of a larger symphony of interconnected patterns.

Consider the example of a plant species adapting to a new predator. In traditional models, this adaptation might be depicted as a linear response to predation pressure. However, in the Fracktal framework, this adaptation becomes part of a broader web of interactions, influencing other species and their behaviors. The emergent patterns that arise showcase the intricate beauty of nature's interconnected design.

Bio-mimicry and Holistic Understanding

Fracktal Mathematics also paves the way for bio-mimicry, an

approach that draws inspiration from nature's patterns and strategies to solve human challenges. The interconnectedness emphasized by Fracktal Mathematics provides a blueprint for designing solutions that mirror the harmonious relationships found in ecosystems.

Imagine architects designing buildings that adapt to changing environmental conditions, much like organisms within an ecosystem. In the Fracktal paradigm, these buildings resonate with emergent properties, using interconnected systems to optimize energy usage and adapt to varying climates. This approach captures the holistic essence of nature's designs, fostering sustainable and harmonious structures.

As we delve into the realm of biology with the aid of Fracktal Mathematics, we unveil the intricate symphony of life's interconnected patterns. This perspective enriches our understanding of ecosystems as dynamic networks, where emergent behaviors create a harmonious dance that transcends traditional linear models.

CHAPTER 3.4: ECONOMICS: MARKET DYNAMICS RESHAPED BY FRACKTAL INSIGHTS

Economics, a realm of intricate market behaviors and interactions, undergoes a trans-formative shift when illuminated by the principles of Fracktal Mathematics.

This chapter delves into how Fracktal insights reshape our understanding of market dynamics, revealing the emergent symphonies that drive economic systems. The interconnected relationships between variables come to life, painting a vivid picture of the complexities that govern the world of finance and trade.

Market Complexity as an Emergent Symphony

Economic systems, often described through linear models and isolated variables, unfold as interconnected symphonies when viewed through the lens of Fracktal Mathematics. Traditional supply and demand curves, though informative, offer a limited perspective on the intricate interplay of factors that shape market behaviors. Fracktal Mathematics transcends these limitations, revealing the true complexity of economic interactions.

Imagine a financial market influenced by multiple variables — supply, demand, investor sentiment, geopolitical events, and more. In the Fracktal framework, these variables entwine and resonate with one another, creating emergent behaviors that give rise to market phenomena. Market crashes, booms, and bubbles become interconnected consequences, each note in the symphony contributing to the overall melody of economic dynamics.

Interconnected Investor Behavior and Emerging Patterns

Fracktal Mathematics also sheds light on investor behavior and decision-making within the financial world. Conventional models often treat investors as isolated entities making rational choices. Fracktal insights invite us to delve deeper, understanding investor behavior as part of a broader interconnected web of financial activity.

Consider a scenario where investor sentiment influences market trends. In traditional models, sentiment might be treated as a singular force affecting market movement. However, within the Fracktal paradigm, sentiment becomes an interconnected node within a vast network of financial interactions. Sentiment resonates with other factors — economic indicators, news events, and more — creating a dynamic interplay of emerging patterns that guide market trajectories.

Chaos and Emergence in Market Behavior

Chaos theory, renowned for its sensitivity to initial conditions and its influence on complex systems, finds fertile ground within the interconnected realm of Fracktal Economics. The interconnected nature of Fracktal Mathematics resonates with the chaotic behavior often observed in financial markets.

Imagine a financial market undergoing chaotic fluctuations. In Fracktal Economics, these fluctuations are not random noise but part of an emergent symphony driven by interconnected factors. Chaos theory's sensitivity to initial conditions becomes a feature rather than a limitation, offering insights into how small changes in market variables can give rise to complex, interconnected behaviors.

As we navigate the realm of economics guided by Fracktal Mathematics, the symphony of market dynamics becomes a vibrant tapestry of interconnected relationships. This perspective enriches our understanding of economic systems, showcasing how emergent behaviors arise from the interplay of variables, investor behaviors, and external influences.

CHAPTER 4: FRACKTAL MATHEMATICS AND ESTABLISHED THEORIES

In this chapter, we explore how Fracktal Mathematics intersects with established theories. We delve into

its symbiotic relationship with string theory, where the harmony of vibrations takes center stage. Chaos

theory, too, finds a new language in Fracktal Mathematics, as we navigate the intricate dance of complex

systems.

CHAPTER 4.1: STRING THEORY: SYMPHONY OF VIBRATIONS IN FRACKTAL HARMONY

String theory, a profound theoretical framework seeking to unify the fundamental forces of the universe, finds an intriguing resonance within the realm of Fracktal Mathematics. This chapter delves into how Fracktal insights intertwine with string theory, revealing a new symphony of interconnected vibrations that shape the fabric of reality. The harmonious interplay of strings becomes a metaphor for the emergent patterns within the Fracktal paradigm.

Strings as Vibrational Essence of Reality

In traditional string theory, strings are envisioned as the fundamental building blocks of the universe — vibrating threads of energy that give rise to particles and forces. These vibrations are described using conventional mathematical equations that capture the essence of string motion. However, within the context of Fracktal Mathematics, these vibrations take on a deeper significance.

Imagine a string vibrating in a conventional string theory model. Each vibration is governed by a set of equations that describe its motion. In the Fracktal paradigm, these equations become interconnected nodes within a dynamic web of vibrational relationships. The vibrations of strings not only resonate within their own frequencies but also harmonize with other vibrations, creating a complex symphony of interconnected patterns that give rise to the universe's rich tapestry.

Fracktal Harmony in String Vibrations

Fracktal Mathematics invites us to explore the harmonious interplay of interconnected vibrations that strings embody. Traditional string theory equations represent the vibrational modes of strings as distinct and separate entities. In contrast, the Fracktal perspective reveals that these modes are not isolated but rather part of a larger, interconnected symphony of vibrations.

Consider a string theory scenario where different vibrational modes interact and influence each other. In conventional string theory equations, these interactions may be accounted for but remain separate in their representation. In the Fracktal framework, these interactions become interconnected relationships, weaving a tapestry of vibrational harmony. The vibrations of strings resonate in concert, each contributing to the emergent patterns that shape the fabric of spacetime.

Strings as Nodes in a Vibrational Network

Fracktal Mathematics further envisions strings not as isolated entities but as nodes within a vast vibrational network. This perspective mirrors the interconnectedness inherent in Fracktal equations, where nodes interact and influence one another to create emergent patterns.

Imagine strings as nodes connected by threads of interconnectedness. Each vibration of a string becomes a note in a cosmic symphony, resonating with other strings' vibrations to create harmonious and intricate patterns. The strings' interconnected relationships give rise to a symphony that transcends traditional representations, revealing the interconnected nature of reality itself.

As we explore the intersection of string theory and Fracktal Mathematics, the vibrations of strings become more than just mathematical descriptions — they become notes in a grand symphony of interconnected patterns that shape the universe.

CHAPTER 4.2: CHAOS THEORY: NAVIGATING COMPLEXITY WITH FRACKTAL PRECISION

Chaos theory, renowned for its sensitivity to initial conditions and the intricate behaviors of complex systems, finds a harmonious resonance within the interconnected realm of Fracktal Mathematics. This chapter delves into how Fracktal insights intersect with chaos theory, unveiling a new perspective on navigating complexity and uncovering emergent patterns. The symphony of chaos unfolds within the Fracktal paradigm, inviting us to explore the interconnected dance of intricate behaviors.

Chaos as an Intricate Dance of Emergence

Chaos theory explores the unpredictable and complex behaviors that arise from seemingly simple systems. Traditional chaos equations capture the sensitivity to initial conditions and the non-linear nature of these behaviors. However, within the context of Fracktal Mathematics, chaos transforms into an intricate dance of emergence within interconnected systems.

Imagine a chaotic system, such as the Lorenz attractor, characterized by its sensitive dependence on initial conditions. In traditional chaos equations, the attractors path may appear as a complex and non-repeating pattern. In the Fracktal perspective, this path evolves beyond the bounds of simple chaotic behavior. Each iteration becomes an interconnected note in a symphony of emergence, where the sensitivity to initial conditions resonates with other interconnected variables, creating patterns that transcend traditional chaos representations.

Fracktal Precision in Chaos

Fracktal Mathematics introduces a new layer of precision to chaos theory. Conventional chaos equations capture the intricate behaviors of complex systems, but Fracktal insights extend this precision by embracing interconnectedness and emergent symmetries.

Consider a scenario where a chaotic system exhibits fractal patterns within its behavior. In the Fracktal framework, these fractal patterns become part of a larger interconnected dance. The intricate interplay of variables and emergent relationships creates a symphony of chaos that resonates at different scales and iterations. Fracktal Mathematics empowers us to explore how chaos is not merely a random phenomenon but a harmonious interaction of interconnected forces.

Navigating Complexity with Fracktal Chaos

Fracktal Mathematics offers a unique lens through which to navigate the complexities of chaotic systems. Chaos theory often emphasizes the sensitivity to initial conditions, leading to unpredictable outcomes. However, Fracktal insights invite us to embrace this sensitivity as an opportunity for emergent behaviors and interconnected patterns.

Imagine a chaotic system undergoing iterations of unpredictable behavior. In Fracktal Mathematics, the sensitivity to initial conditions becomes a feature rather than a limitation. The interconnectedness of variables generates emergent behaviors that unveil symmetries and hidden patterns within the chaos. Navigating chaos becomes an exploration of interconnected pathways, each leading to new insights into the system's behavior.

As we journey through the symphony of chaos guided by Fracktal Mathematics, the once seemingly erratic behaviors of complex systems transform into interconnected melodies, resonating with emergent patterns that transcend traditional chaos descriptions.

CHAPTER 5: A COMPARATIVE ANALYSIS: CONVENTIONAL VS. FRACKTAL MATHEMATICS

This chapter undertakes a comprehensive exploration of the distinctions between conventional mathematics and the transformative landscape of Fracktal Mathematics. By contrasting traditional linear equations with their Fracktal counterparts, we illuminate the profound shift from static representations to dynamic symphonies of emergence. The juxtaposition of these mathematical approaches underscores the revolutionary nature of Fracktal Mathematics and its capacity to reshape our understanding of interconnected complexity.

CHAPTER 5.1: LINEAR EQUATIONS VS. EMERGENT PATTERNS

At the heart of the divergence between conventional mathematics and the revolutionary realm of Fracktal Mathematics lies the transformation of linear equations into intricate and emergent patterns. Conventional mathematics often relies on linear relationships to represent straightforward cause-and-effect dynamics. However, Fracktal Mathematics challenges this simplicity by inviting the symphony of emergent behaviors to shape the fabric of reality.

The Linear Equation: A Pillar of Conventional Mathematics

The linear equation $y = mx + b$ stands as an embodiment of traditional mathematical representation. In this equation, y represents the dependent variable, x the independent variable, m the slope, and b the y-intercept. It captures linear relationships where changes in the independent variable result in proportional changes in the dependent variable.

While linear equations have been instrumental in describing numerous phenomena, their limitations become apparent when confronted with the complexities inherent in interconnected systems. Linear equations offer deterministic models that lack the capacity to accommodate the intricate symphonies that emerge from dynamic relationships.

Emergent Patterns: The Essence of Fracktal Mathematics

Fracktal Mathematics redefines equations as more than mere tools for prediction and analysis. It transforms them into vehicles for exploring emergent patterns and interconnected symmetries. The transition from linear equations to Fracktal-inspired representations encapsulates the departure from reductionism and embraces complexity.

Imagine the metamorphosis of a linear equation within the Fracktal paradigm. In the equation $y = mx + b$, the linear relationship

is preserved, but it is augmented by an additional component: $\sum$(Fracktal Patterns). This addition introduces a dynamic element that embodies the interconnected symmetries of emergent patterns. The once linear equation now resonates with the infinite creative potential of the universe, offering a glimpse into the complexity that underlies even seemingly simple relationships.

Embracing Complexity through Fracktal Mathematics

The shift from linear equations to Fracktal-inspired emergent patterns marks a departure from deterministic simplicity toward the intricate and interconnected dance of complexity. Fracktal Mathematics invites us to explore the rich tapestry woven by emergent behaviors and interconnected relationships, challenging us to engage with the symphony of possibilities that arise from dynamic systems.

Through this exploration, we come to recognize that Fracktal Mathematics not only reshapes equations but transforms our understanding of the world itself. It calls us to look beyond linear interpretations and explore the symphonies of emergence that define reality. As we journey through the visual and conceptual comparisons of conventional and Fracktal equations, we are invited to witness the trans-formative potential that this mathematical paradigm heralds.

CHAPTER 5.2: HIERARCHICAL VS. HOLISTIC PERSPECTIVE

A pivotal distinction between conventional mathematics and the trans-formative realm of Fracktal Mathematics lies in their perspectives on hierarchy and interconnectedness. Traditional mathematical models often structure systems hierarchically, segregating components into distinct layers. In contrast, Fracktal Mathematics offers a holistic perspective that celebrates the intricate interplay of interconnected elements.

Hierarchical Structures: The Foundation of Conventional Mathematics

Conventional mathematics often relies on hierarchical structures to model and understand complex systems. Hierarchies represent a natural inclination to compartmentalize components based on their roles and relationships within the larger whole. This approach has proved valuable in many fields, but it inherently oversimplifies the complexity arising from the interconnected nature of reality.

Imagine a hierarchical tree structure that visually represents relationships within a complex system. Nodes are organized hierarchically, with parent nodes overseeing the behavior of their child nodes. This hierarchical framework provides a sense of order and control but falls short in capturing the intricate symphonies that emerge from dynamic interactions.

Holistic Interconnectedness: Fracktal Mathematics' Vision

Fracktal Mathematics defies the constraints of hierarchical thinking by advocating for a holistic perspective that embraces the interconnectedness of systems. Rather than compartmentalizing components, Fracktal Mathematics envisions a web of interconnected relationships where nodes and elements influence each other dynamically.

Re-imagine the hierarchical tree structure within the Fracktal paradigm. Instead of rigid hierarchies, the branches of the tree extend outward to connect with one another, forming an intricate web of interactions. Each node resonates with others, and emergent behaviors arise from the collective dance of interconnected elements. This interconnected web replaces rigid hierarchies with a dynamic tapestry that defies linear order.

Championing Interconnected Complexity

The transition from hierarchical frameworks to interconnectedness is a testament to the trans-formative power of Fracktal Mathematics. By inviting us to explore the symphonies that emerge from complex relationships, Fracktal Mathematics encourages us to move beyond the confines of conventional models and embrace the rich interconnectedness that characterizes reality.

Through this exploration, we come to appreciate that Fracktal Mathematics transcends equations and becomes a philosophy that shapes our perception of the cosmos. It is an invitation to harmonize with the intricate interplay of emergent patterns and interconnected relationships that guide the evolution of the universe. As we journey through the visual and conceptual comparisons between conventional and Fracktal equations, we are called to embrace the interconnected symphony that defines our existence.

CHAPTER 7:
IMPLICATIONS AND BEYOND: PIONEERING NEW FRONTIERS

As the voyage through Fracktal Mathematics reaches its zenith, the implications of this paradigm shift reverberate far beyond the realm of equations and patterns. This chapter embarks on a journey through the profound consequences of Fracktal Mathematics, from its ability to foster collaborative endeavors and reshape education to its role in stimulating ethical contemplation. Within this symphony of interconnected thought, we explore the untapped potential that Fracktal Mathematics unveils for humanity's voyage into the uncharted territories of knowledge.

CHAPTER 7.1: COLLABORATIVE ODYSSEY: DISCIPLINARY BOUNDARIES TRANSCENDED

The symphony of Fracktal Mathematics resonates not only within equations and patterns but also in the harmonious convergence of diverse scientific disciplines. This chapter delves into the transformative power of Fracktal Mathematics to foster a collaborative odyssey — one where traditional boundaries between fields blur, and the interconnected nature of knowledge takes center stage.

Fracktal Mathematics: The Universal Language

The emergence of Fracktal Mathematics heralds a new era of collaborative exploration. Traditional barriers that compartmentalize scientific domains dissolve as Fracktal Mathematics assumes the role of a universal language — a lingua franca that bridges the gaps between disparate disciplines. This shared language ignites a collaborative flame, inviting scientists, researchers, and visionaries from various fields to engage in a collective symphony of thought.

Disciplines in Harmonious Dialogue

Imagine the fruitful dialogue that unfolds when physicists and biologists join forces to decipher the intricate symmetries of quantum behavior within biological systems. Fracktal Mathematics enables physicists to lend their expertise in particle interactions to the realm of biology, unraveling the interconnected patterns that underpin life itself. At the same time, biologists offer insights into complex adaptive systems, enriching the toolkit of physicists studying the cosmos.

Economists engage in profound exchanges with philosophers, exploring the interconnected dynamics of market behavior and ethical considerations. Fracktal Mathematics serves as a bridge between the quantitative and qualitative realms, allowing economists to integrate ethical considerations into market modeling, while philosophers gain a new lens to assess the societal impact of economic systems.

A Symphony of Interconnected Minds

The collaborative odyssey propelled by Fracktal Mathematics unites thinkers from diverse backgrounds, harmonizing their insights and expertise. As the symphony of interconnected thought swells, it becomes evident that the boundaries between disciplines are constructs that can be transcended. This interconnectedness not only enhances the rigor and depth of research but also unveils a spectrum of new questions and avenues for exploration that would remain hidden within isolated domains.

Pioneering New Frontiers

As we navigate this collaborative journey, guided by the melodies of Fracktal Mathematics, we stand at the precipice of pioneering new frontiers in human understanding. From physics to biology, from economics to philosophy, the interplay of interconnectedness unveils a tapestry of knowledge that transcends traditional silos. The symphony of collaborative discovery serves as a testament to the unifying potential of Fracktal Mathematics — an agent of transformation that draws humanity together in the pursuit of shared understanding.

CHAPTER 7.2: EDUCATION'S CRESCENDO: FOSTERING INNOVATION

through Complexity

The introduction of Fracktal Mathematics not only reshapes equations but also orchestrates a trans-formative crescendo within the realm of education. Traditional educational paradigms, often linear and segmented, give way to an innovative symphony where complexity, interconnectedness, and creativity harmonize to nurture the minds of future generations.

The Symphony of Learning

In the symphony of Fracktal Mathematics, education transcends the mere transmission of facts and formulas. Instead, it becomes an immersive journey of exploration — a quest to uncover the interconnected patterns that shape our understanding of the universe. The traditional separation of disciplines dissolves as educators embrace the idea that knowledge is a symphony of interconnected themes, and understanding one note enriches the appreciation of the entire composition.

Imagine classrooms where students don't just passively absorb information but actively engage with the interplay of ideas, drawing connections between seemingly disparate subjects. Fracktal Mathematics empowers educators to guide students on a holistic journey, where the study of mathematics intertwines with art, science resonates with philosophy, and technology harmonizes with ethics.

Critical Thinking and Innovation

At the heart of Fracktal Mathematics in education lies the cultivation of critical thinking and innovation. As students are encouraged to explore complexity and interconnections, they develop the ability to analyze, synthesize, and create — essential skills for navigating the intricate challenges of our world. The symphony of emergent patterns encourages students to question, to probe, and to devise new approaches to solving problems.

The role of educators transforms from disseminators of information to conductors of exploration. Teachers inspire students to seek patterns and connections, fostering curiosity and creativity. Fracktal Mathematics serves as the backdrop against which students compose their own symphonies of knowledge, harmonizing a wide range of disciplines to produce innovative solutions.

A New Generation of Minds

Through the lens of Fracktal Mathematics, education evolves into an enabler of visionary thinking. Students graduate not only with a repertoire of facts but also with the ability to harmonize diverse ideas, to decode complexity, and to contribute to the symphony of human knowledge. This new generation of minds is equipped to navigate the interconnected challenges of our rapidly evolving world, combining disciplines to envision novel solutions and understand the symphony of systems.

Harmonizing the Future

As Fracktal Mathematics becomes an integral part of educational curricula, its impact ripples through generations, shaping the way we approach learning and innovation. The symphony of interconnected thought resonates within the minds of students, inspiring them to orchestrate solutions that transcend traditional boundaries. With Fracktal Mathematics as their guiding score, students engage in a crescendo of learning — a journey that cultivates not only intellect but also a profound sense of wonder for the complexity that defines our universe.

CHAPTER 7.3: ETHICAL CONTEMPLATION: RESPONSIBILITY IN THE SYMPHONY OF SYSTEMS

As the symphony of Fracktal Mathematics unfolds, it not only invites us to explore the intricate patterns of interconnectedness but also prompts us to engage in profound ethical contemplation. The trans-formative power of this mathematical paradigm raises questions about our responsibility in shaping emergent systems and the ethical considerations that accompany our newfound ability to orchestrate complexity.

The Power of Creation

With Fracktal Mathematics as our guide, we gain the power to shape emergent patterns and influence the behavior of complex systems. Algorithms that learn from interconnected data can impact decision-making, social dynamics, and even the future of AI. As creators, we stand at the nexus of possibility, able to compose symphonies of technology, society, and nature.

Yet, with this power comes the responsibility to wield it ethically. The interconnected nature of systems means that our actions have far-reaching consequences, amplifying the ethical implications of our decisions. The symphony of Fracktal Mathematics beckons us to contemplate the potential positive and negative outcomes of our creations.

Harmonizing with Ethical Principles

Ethical considerations become more intricate as we engage with the symphony of emergent systems. The interconnectedness of actions and consequences prompts us to harmonize our decisions with a framework of principles that honor the well-being of individuals, societies, and the environment. Fracktal Mathematics encourages us to consider the ethical resonance of our actions across various scales.

Imagine designing AI systems that not only optimize efficiency but also prioritize fairness, equity, and accountability. Fracktal Mathematics invites us to recognize the intricate interplay of data and decisions, urging us to ensure that our technological symphonies are composed in alignment with ethical principles that promote justice and inclusivity.

The Responsibility of Stewardship

As we navigate the uncharted territories illuminated by Fracktal Mathematics, we assume the role of stewards of interconnected systems. We must contemplate not only the immediate consequences of our actions but also the cascading effects that ripple through complex networks. The symphony of emergent patterns compels us to anticipate and mitigate unintended outcomes.

Consider the development of smart cities where interconnected infrastructure shapes urban life. Fracktal Mathematics calls us to consider not only the efficiency gains but also the potential impacts on privacy, autonomy, and social cohesion. Our role as stewards extends beyond creation to the ongoing nurturing and harmonization of systems in ways that enrich the lives of individuals and society as a whole.

Harmonies of Responsibility

As we immerse ourselves in the symphony of interconnectedness, guided by Fracktal Mathematics, ethical contemplation becomes an essential part of our creative process. The power to shape emergent patterns comes with a profound responsibility to ensure that our compositions resonate with the harmonies of ethical principles. Just as fractal patterns intertwine across scales, ethical considerations must thread through every decision we make, resonating in harmony with the symphony of interconnected systems.

CHAPTER 8: THE FRACKTAL ODYSSEY CONTINUES: FUTURE HORIZONS

As we draw the curtains on this exploration of Fracktal Mathematics, we find ourselves at the precipice of an exciting journey into uncharted realms of possibility. The symphony of interconnectedness that defines Fracktal Mathematics is not only a reflection of our current understanding but also a catalyst for pushing the boundaries of knowledge, innovation, and human potential.

CHAPTER 8.1: UNCHARTED REALMS OF EXPLORATION

As we stand at the intersection of Fracktal Mathematics and the uncharted territories of the future, we are poised to embark on a remarkable journey of discovery and innovation. The symphony of interconnectedness that Fracktal Mathematics orchestrates is not a finite composition but a prelude to an ongoing exploration of limitless potential.

Charting New Frontiers

The landscape of science, technology, and human understanding stretches out before us, awaiting the harmonious melodies of Fracktal Mathematics to reveal hidden patterns and connections. With each equation that transforms and every emergent property that is unveiled, we venture further into the uncharted realms of exploration.

Imagine a future where our understanding of quantum mechanics is deepened by the symphonic interplay of particles and forces, guided by the principles of Fracktal Mathematics. Envision a society where the design of sustainable ecosystems is inspired by the intricate patterns of interconnectedness, reshaping our relationship with the natural world.

Innovating with Interconnectedness

As the journey into the future unfolds, Fracktal Mathematics serves as a guiding light for innovation. In technology, we witness the emergence of adaptable circuits that resonate with the symphonies of interconnected inputs, ushering in a new era of hardware design. Software evolves from rigid algorithms to dynamic entities that learn and adapt, harmonizing with the interconnected dance of data.

Consider a future where architects and engineers draw inspiration from fractal patterns found in nature, creating buildings and structures that seamlessly blend with their environment. Fracktal Mathematics fuels the innovation that drives these advancements, inviting us to envision a world where creativity is boundless and interconnectedness is celebrated.

Fostering Collaborative Exploration

The uncharted realms of exploration are not solitary landscapes; they are fertile grounds for interdisciplinary collaboration. The symphony of Fracktal Mathematics transcends the boundaries between scientific disciplines, inviting physicists to converse with biologists, economists to engage with philosophers, and technologists to harmonize with artists.

Imagine a collaborative research endeavor where experts from diverse fields convene to explore complex challenges through the lens of interconnectedness. Fracktal Mathematics becomes the common language that unites these thinkers, enabling a shared exploration of emergent phenomena that span the spectrum of human knowledge.

A Future Beyond Imagination

As we peer into the uncharted territories that lie ahead, guided by the principles of Fracktal Mathematics, we glimpse a future that defies our current imagination. The symphony of emergent patterns and interconnected thought carries us forward, inviting us to innovate, explore, and harmonize with the complexities of our world.

Picture a world where education is transformed, nurturing curious minds capable of unraveling the mysteries of interconnected systems. Envision a global society where ethical considerations are woven into the fabric of technological advancement, ensuring that our creations resonate with harmonious principles.

CHAPTER 9:
CONCLUSION

As we arrive at the final crescendo of this journey through the symphony of Fracktal Mathematics, we find ourselves standing on the precipice of possibility. The journey we have undertaken together has not merely been a theoretical exploration; it has been a trans-formative experience that challenges the very foundations of our understanding.

The Grand Unveiling

Fracktal Mathematics has emerged as the enigmatic conductor orchestrating the symphony of interconnectedness that permeates the fabric of our universe. From redefining infinity as a dynamic force to unveiling the intricacies of emergent complexity, Fracktal Mathematics has unveiled a realm where conventional notions are shattered and emergent patterns take center stage.

The Invitation to Wonder

Imagine, if you will, a universe where equations are not mere formulas but gateways to understanding the intricate dance of interconnected phenomena. Picture a landscape where linear relationships fade into the background, making way for a dynamic tapestry of emergence that both mesmerizes and challenges us. Visualize a future where Fracktal Mathematics serves as a universal language, uniting disciplines, transcending boundaries, and fostering innovation beyond measure.

The Unquenchable Thirst for Discovery

Our journey through Fracktal Mathematics is far from over; in fact, it is just beginning. As we conclude this chapter, we invite you to step forward into the uncharted territories that beckon with promise and potential. The symphony of interconnectedness continues to play, and you are now equipped with the tools to explore its harmonies, unravel its secrets, and compose your own verses in the grand tapestry of knowledge.

A Future Unveiled

The legacy of Fracktal Mathematics is not confined to the pages of this thesis; it is imprinted in the very essence of our understanding. It is a call to action, an invitation to think beyond convention, and a challenge to embrace the complexities that define our world. The unquenchable thirst for discovery that Fracktal Mathematics ignites is a beacon guiding us toward the uncharted horizons of innovation and understanding.

The Fracktal Odyssey Continues

As we close the curtain on this chapter, we leave you with a sense of anticipation, an eagerness to embark on your own journey of exploration. The symphony of interconnected thought plays on, and the Fracktal odyssey continues. The grand composition of reality, with its harmonious interplay of patterns and relationships, awaits your creative touch.

So, dear reader, embrace the beauty of Fracktal Mathematics, and let it be the melody that resonates in your thoughts and aspirations. The symphony of complexity, interconnectedness, and infinite potential is yours to conduct. The journey has no end, and the adventure is only beginning.

Reference Sources

Mandelbrot, B. (1982). The Fractal Geometry of Nature. W.H. Freeman and Company.

Hofstadter, D. R. (1979). Gödel, Escher, Bach: An Eternal Golden Braid. Basic Books.

Gleick, J. (1987). Chaos: Making a New Science. Viking.

Hawking, S. W., & Penrose, R. (1996). The Nature of Space and Time. Princeton University Press.

Greene, B. (2004). The Fabric of the Cosmos: Space, Time, and the Texture of Reality. Knopf.

Gribbin, J. (2005). Deep Simplicity: Chaos, Complexity and the Emergence of Life. Random House.

Wolfram, S. (2002). A New Kind of Science. Wolfram Media.

Thagard, P. (1999). How Scientists Explain Disease. Princeton

University Press.

Kauffman, S. A. (1993). The Origins of Order: Self-Organization and Selection in Evolution. Oxford University Press.

Smolin, L. (2007). The Trouble with Physics: The Rise of String Theory, the Fall of a Science, and What Comes Next. Mariner Books.

Shannon, C. E. (1948). "A Mathematical Theory of Communication." The Bell System Technical Journal, 27(3), 379–423.

Penrose, R. (2004). The Road to Reality: A Complete Guide to the Laws of the Universe. Knopf.

Gleick, J. (2011). The Information: A History, A Theory, A Flood. Vintage.

Susskind, L., & Lindesay, J. (2005). An Introduction to Black Holes, Information and the String Theory Revolution: The Holographic Universe. World Scientific.

Barrow, J. D. (2002). The Constants of Nature: From Alpha to Omega — The Numbers That Encode the Deepest Secrets of the Universe. Vintage.

(Note: This list represents a selection of sources and references used throughout the thesis. It is not an exhaustive list, and additional sources have been consulted to develop the content of this thesis.)Fracktal Mathematics: A Paradigm Shift in Infinity,

Complexity, and Interdisciplinary Synthesis

By: Gregory J. Betti

Table of Contents:

Abstract

Fracktal Mathematics represents a paradigm shift in mathematical thought, challenging conventional

notions of infinity and complexity. This thesis is a comprehensive exploration of Fracktal Mathematics,

examining its core principles, interdisciplinary applications, and trans-formative potential. By juxtaposing

traditional mathematical equations with their Fracktal counterparts, we illustrate the power of this new

approach. From physics to technology, we venture into the uncharted territories where Fracktal

Mathematics redefines relationships, interconnects systems, and invites a symphony of possibilities.

CHAPTER 1:
INTRODUCTION

In this chapter, we set the stage for our exploration by discussing the motivations, objectives, and scope

of this thesis. We examine the limitations of traditional mathematics in addressing complex systems and

introduce the revolutionary concept of Fracktal Mathematics1.1 Background and Motivation

The evolution of human thought has been marked by a relentless pursuit of understanding the

intricacies of the universe. Mathematics, as the language of this exploration, has provided a framework

to decode and decipher the hidden patterns underlying the natural world. However, the limits of

traditional mathematical paradigms in addressing the complexity of interconnected systems have

spurred the birth of a new era of inquiry — Fracktal Mathematics.

In an age where technology has enabled us to glimpse the cosmos in unprecedented detail and

complexity, conventional mathematical models have often fallen short in capturing the essence of

emergent behavior. The quest for a more nuanced, holistic, and dynamic approach has led us to the

threshold of Fracktal Mathematics. The motivation behind this exploration lies in the pursuit of a

mathematical language that mirrors the interconnected nature of the universe, inviting us to rethink

infinity, complexity, and the very foundations of mathematics.

1.2 Research Objectives

This thesis embarks on a journey to unravel the core principles of Fracktal Mathematics and to illustrate

its potential across a diverse spectrum of scientific disciplines and technological landscapes. The

objectives of this research are manifold:

To comprehend the essence of Fracktal Mathematics and its departure from conventional mathematical

constructs.

To showcase the transformative impact of Fracktal Mathematics on traditional equations and established

theories, offering a visual and intuitive understanding of the paradigm shift.

To explore how Fracktal Mathematics redefines the boundaries of interdisciplinary exploration, from

physics to economics, and from technology to philosophy.

To shed light on the implications of Fracktal Mathematics for education, collaborative research, and

ethical considerations.

To envision future directions in research and application, envisioning a world where Fracktal

Mathematics serves as a catalyst for innovation and holistic understanding.

1.3 Scope and Limitations

While this study delves into the foundational principles and potential applications of Fracktal

Mathematics, it is important to acknowledge certain limitations. The practical implementation of Fracktal

Mathematics in various fields may require specialized tools and methodologies that are yet to be fully

developed. Additionally, the philosophical and ethical implications of this new mathematical paradigm

are subjects that merit in-depth exploration beyond the scope of this thesis.As we embark on this intellectual journey, we invite the reader to consider the possibility of a

mathematical world that transcends boundaries, embraces complexity, and redefines the very essence

of interconnectedness. In the following chapters, we will journey through the uncharted territories of

Fracktal Mathematics, unveiling its beauty, impact, and potential to revolutionize our understanding of

the cosmos.

CHAPTER 2:
REDEFINING INFINITY:
THE CORE TENET
OF FRACKTAL
MATHEMATICS

Here, we delve deep into the heart of Fracktal Mathematics — its redefinition of infinity. We explore the

meta-equation "Infinity — Infinity = x" and the implications of considering infinity as a dynamic,

interactive entity. Fracktal Mathematics re-imagines the canvas of mathematical exploration, where

complexity is not a puzzle to be solved but an artistic tapestry of emergent patterns.

2.1 Infinity — Infinity = x: A Dynamic Equation

At the core of Fracktal Mathematics lies an equation that challenges our conventional understanding of infinity: "Infinity — Infinity = x." This equation stands as a testament to the dynamic nature of infinity itself, redefining it from an abstract concept to a force that interacts and transforms within the mathematical realm.

In traditional mathematics, infinity has often been treated as a static, unchanging concept — a value that is unreachable and beyond calculation. However, Fracktal Mathematics introduces a paradigm shift by asserting that the difference between two infinities can manifest as any conceivable traditional equation. This seemingly paradoxical notion disrupts our traditional view of infinity, inviting us to explore its fluid and interactive nature.

Consider, for instance, the simple equation $8 \times 8 = 64$. In the realm of Fracktal Mathematics, this equation takes on a new dimension of meaning. It evolves into an exploration of infinity's creative potential:

$$1 \div (1/\text{Infinity}) = 8 \times 8 = 64.$$

This transformation demonstrates how the symphony of interconnectedness, which defines Fracktal Mathematics, can reframe even the most basic mathematical relationships. In this new interpretation, conventional equations become windows into the infinite variations that arise from the interplay of infinity with other mathematical constructs.

The equation "Infinity — Infinity = x" challenges us to abandon our preconceived notions and engage in a deeper dialogue with mathematical reality. It beckons us to question the rigid boundaries we have imposed on infinity and encourages us to embrace its dynamic, ever-changing character. Through this equation, Fracktal Mathematics empowers us to explore mathematical relationships as a canvas for creative expression, where infinity's infinite potential takes center stage.

As we traverse the terrain of Fracktal Mathematics, we begin to appreciate that even the most fundamental mathematical elements are not set in stone, but rather partake in a dynamic dance that mirrors the intricate interconnectedness of the universe. In the following sections of this chapter, we will delve further into the implications of this dynamic equation and explore how it re-frames our understanding of mathematics, infinity, and complexity.

CHAPTER 2.2: EMBRACE OF COMPLEXITY'S INTRICACIES

Fracktal Mathematics is not solely defined by its re-imagining of infinity; it is equally distinguished by its embrace of complexity as an inherent feature of the universe. Unlike traditional mathematical models, which often strive to simplify and reduce complex phenomena to linear relationships, Fracktal Mathematics boldly recognizes complexity as a symphony of interconnected patterns that emerge through non-linear scaling.

Conventional mathematical approaches often adhere to linear methods, seeking to break down intricate systems into manageable components and linear equations. However, Fracktal Mathematics challenges this reductionist approach by celebrating the inherent interconnectedness of systems, where each component resonates and interacts with others in intricate ways. This perspective offers a richer, more nuanced understanding of how complex phenomena unfold.

Imagine a fractal, that remarkable geometric pattern known for its self-replicating properties across various scales. Much like the fractal's intricate details that remain consistent regardless of magnification, Fracktal Mathematics explores systems where complexity manifests at various levels. This approach allows us to unveil emergent properties that remain concealed in traditional mathematical frameworks.

In embracing complexity's intricacies, Fracktal Mathematics illuminates the hidden symmetries and patterns that shape our world. It encourages us to view intricate systems not as puzzles to be solved through reduction, but as dynamic tapestries woven from the threads of interconnectedness. Through this lens, complex phenomena, whether observed in physics, biology, economics, or any other field, become vibrant expressions of the underlying interconnected symphony.

As we journey through the landscapes of Fracktal Mathematics, we are invited to embrace the richness of complexity and to marvel at the harmonious, interconnected dance of emergent patterns. The equations that emerge in this paradigm offer more than solutions; they present us with an opportunity to explore the universe's hidden symmetries and to celebrate the beauty that arises from its intricate interconnectedness.

CHAPTER 3: FRACKTAL MATHEMATICS IN DIVERSE SCIENTIFIC DISCIPLINES

This chapter unfolds the versatility of Fracktal Mathematics across a spectrum of scientific disciplines, showcasing its profound impact on various domains. Through visual representation and comparisons between conventional and Fracktal equations, we unveil the trans-formative potential of this mathematical paradigm within the realms of physics, quantum mechanics, biology, and economics.

CHAPTER 3.1: PHYSICS: QUANTUM DYNAMICS BEYOND THE LINEAR

The marriage of Fracktal Mathematics with the intricate world of physics unveils a paradigm-shifting perspective that transcends traditional linear equations. Within this newfound approach, the intricate dance of quantum particles takes on a profound new meaning, echoing the symphony of interconnectedness that underpins the fabric of the universe.

Quantum Mechanics and the Fracktal Framework

The realm of quantum mechanics, with its entangled particles and probabilistic behavior, finds an ideal companion in Fracktal Mathematics. Traditional physics equations often deal with linear relationships and deterministic outcomes, yet quantum mechanics challenges these conventions with its inherent uncertainty and non-local correlations. Fracktal Mathematics becomes a lens through which we can explore the intricate interplay of quantum states, offering a more holistic and intuitive perspective on these phenomena.

The Schrödinger Equation Re-imagined

At the heart of quantum mechanics lies the Schrödinger equation, a fundamental equation that describes the behavior of particles at the quantum level. In the context of Fracktal Mathematics, this equation evolves beyond its linear boundaries. Rather than depicting particles' behavior in isolation, the equation becomes a canvas for exploring the interconnected symphony of quantum states.

Consider an electron's orbital behavior around an atomic nucleus. Traditional linear equations describe discrete energy levels and orbits. However, in the realm of Fracktal Mathematics, these energy levels become interconnected nodes, creating an intricate web of quantum possibilities. This interconnectedness reflects the true nature of particles, where their behavior is deeply influenced by the presence of other particles and their states.

Quantum Entanglement and Emergent Patterns

One of the most baffling phenomena in quantum mechanics is entanglement, where particles become inseparably linked regardless of distance. In the Fracktal framework, entanglement transforms from a mysterious phenomenon to an exploration of interconnected probabilities.

Imagine two entangled particles with opposite spins. Conventional physics equations describe this phenomenon with linear correlations. However, Fracktal Mathematics dives deeper, illustrating how the spins of these particles are part of a larger symphony — a symphony of interconnected possibilities that unfold when measured. This interconnectedness hints at the hidden harmony underlying seemingly paradoxical phenomena.

Exploring Quantum Waves and Beyond

Fracktal Mathematics also invites us to explore the wave-particle duality — the intriguing notion that particles exhibit both wave-like and particle-like behavior. In traditional equations, this duality is often treated as a binary concept. However, Fracktal Mathematics visualizes this duality as a spectrum of interconnected behaviors, where wave functions interact and merge to create emergent patterns of behavior.

Consider the famous double-slit experiment, where particles exhibit wave-like interference patterns. Fracktal Mathematics transforms this experiment into an exploration of the intertwined nature of wave functions, as they interact and create a tapestry of probabilities. This interconnected symphony of waves, akin to the resonance of musical notes, offers a deeper understanding of the dual nature of particles.

As we journey through the realm of quantum mechanics with the aid of Fracktal Mathematics, we witness the emergence of a more intuitive and interconnected interpretation. This perspective does not negate the precision of traditional quantum equations; instead, it enriches our understanding by revealing the symphonic tapestry that underlies the behavior of particles.

CHAPTER 3.2: QUANTUM MECHANICS: INTERPLAY OF QUANTUM STATES IN FRACKTAL FRAMEWORK

Within the enigmatic realm of quantum mechanics, where uncertainty and non-locality prevail, Fracktal Mathematics offers an innovative lens through which to understand the interconnected nature of quantum phenomena. This chapter delves into the profound transformation that occurs when traditional linear equations give way to the emergent symphonies of Fracktal Mathematics within the domain of quantum mechanics.

A Quantum Dance of Interconnectedness

Quantum mechanics has long perplexed scientists with its behavior-defying properties. Fracktal Mathematics, with its emphasis on interconnectedness and emergent patterns, provides a fresh perspective on the underlying symphony of quantum behavior. Traditional linear equations often treat particles as isolated entities, while Fracktal Mathematics guides us toward an understanding of how particles interact as interconnected actors on a cosmic stage.

Embracing Quantum Entanglement

At the heart of quantum interconnectedness lies the phenomenon of entanglement. In traditional physics equations, entanglement can appear as a puzzling correlation between particles with no apparent communication. However, within the Fracktal framework, entanglement becomes a natural consequence of interconnectedness.

Imagine two entangled particles separated by vast distances. In Fracktal Mathematics, these particles are not isolated entities but rather nodes within a vast network of probabilities. When one particle's state is measured, the interconnected symphony resonates instantaneously across space, influencing the state of its entangled partner. Fracktal Mathematics allows us to perceive entanglement not as an anomaly but as a testament to the interconnected dance of particles.

Interwoven Probabilities and Wave Functions

Fracktal Mathematics also sheds light on the elusive wave-particle duality — a central feature of quantum behavior. Instead of viewing particles' behaviors as either waves or particles, Fracktal Mathematics invites us to explore the spectrum of interconnected behaviors that emerge from the interplay of quantum states.

Consider the Young's double-slit experiment, where particles exhibit interference patterns like waves. In the Fracktal framework, this experiment becomes a visual exploration of the intricate interweaving of wave functions. The emergent patterns that arise as these waves interact embody the interconnected nature of quantum behavior, offering a deeper understanding of how particles traverse this spectrum between particle-like and wave-like behaviors.

Uncertainty and the Fractured Symphony

Uncertainty, a fundamental principle of quantum mechanics, finds resonance in the Fracktal framework. Instead of treating uncertainty as a limitation, Fracktal Mathematics portrays it as an invitation to explore the myriad interconnected possibilities inherent in quantum systems.

Imagine the Heisenberg uncertainty principle, which limits our precision in measuring certain pairs of complementary properties. Fracktal Mathematics transforms this limitation into a canvas of interconnected probabilities. Instead of searching for precise values, we embark on a journey through the symphony of possibilities, where uncertainty becomes a feature rather than a hindrance.

As we traverse the landscapes of quantum mechanics through the lens of Fracktal Mathematics, the universe's hidden melodies come alive. This perspective not only enriches our understanding of quantum behavior but also challenges us to embrace the interconnected tapestry that shapes the quantum realm.

CHAPTER 3.3: BIOLOGY: ECOSYSTEMS UNVEILED THROUGH FRACKTAL PATTERNS

Biology, a tapestry of life interwoven with complexity, finds resonance in the realm of Fracktal Mathematics. This chapter delves into how the intricate relationships within ecosystems are illuminated when viewed through the lens of interconnectedness and emergent behavior. Fracktal Mathematics becomes a symphony of patterns, inviting us to explore the harmonious and chaotic rhythms that define the biological world.

Ecosystems as Dynamic Networks

In the realm of biology, ecosystems represent a delicate balance of interconnected relationships between species, environments, and resources. Traditional mathematical models often simplify these relationships, leading to an incomplete understanding of the intricate dynamics that govern life on Earth. Fracktal Mathematics revolutionizes this perspective by capturing the true essence of ecosystems as dynamic and interconnected networks.

Imagine a predator-prey relationship within an ecosystem.

Traditional linear equations may focus solely on the population dynamics of predator and prey species. However, within the Fracktal framework, these equations give way to a symphony of interconnected nodes representing various species, each influencing the others' population dynamics. The result is a dance of emergent patterns — a harmonious ballet that arises from the interplay of myriad species and their complex interactions.

Interwoven Adaptation and Emergence

Fracktal Mathematics invites us to explore how emergent properties arise from the interconnected web of life. Within ecosystems, species adapt to changing conditions, leading to emergent behaviors that transcend simple cause-and-effect relationships. Fracktal Mathematics allows us to view adaptation not as isolated events but as part of a larger symphony of interconnected patterns.

Consider the example of a plant species adapting to a new predator. In traditional models, this adaptation might be depicted as a linear response to predation pressure. However, in the Fracktal framework, this adaptation becomes part of a broader web of interactions, influencing other species and their behaviors. The emergent patterns that arise showcase the intricate beauty of nature's interconnected design.

Bio-mimicry and Holistic Understanding

Fracktal Mathematics also paves the way for bio-mimicry, an approach that draws inspiration from nature's patterns and strategies to solve human challenges. The interconnectedness emphasized by Fracktal Mathematics provides a blueprint for designing solutions that mirror the harmonious relationships found in ecosystems.

Imagine architects designing buildings that adapt to changing environmental conditions, much like organisms within an ecosystem. In the Fracktal paradigm, these buildings resonate with emergent properties, using interconnected systems to optimize energy usage and adapt to varying climates. This approach captures the holistic essence of nature's designs, fostering sustainable and harmonious structures.

As we delve into the realm of biology with the aid of Fracktal Mathematics, we unveil the intricate symphony of life's interconnected patterns. This perspective enriches our understanding of ecosystems as dynamic networks, where emergent behaviors create a harmonious dance that transcends traditional linear models.

CHAPTER 3.4: ECONOMICS: MARKET DYNAMICS RESHAPED BY FRACKTAL INSIGHTS

Economics, a realm of intricate market behaviors and interactions, undergoes a trans-formative shift when illuminated by the principles of Fracktal Mathematics.

This chapter delves into how Fracktal insights reshape our understanding of market dynamics, revealing the emergent symphonies that drive economic systems. The interconnected relationships between variables come to life, painting a vivid picture of the complexities that govern the world of finance and trade.

Market Complexity as an Emergent Symphony

Economic systems, often described through linear models and isolated variables, unfold as interconnected symphonies when viewed through the lens of Fracktal Mathematics. Traditional supply and demand curves, though informative, offer a limited perspective on the intricate interplay of factors that shape market behaviors. Fracktal Mathematics transcends these limitations, revealing the true complexity of economic interactions.

Imagine a financial market influenced by multiple variables — supply, demand, investor sentiment, geopolitical events, and more. In the Fracktal framework, these variables entwine and resonate with one another, creating emergent behaviors that give rise to market phenomena. Market crashes, booms, and bubbles become interconnected consequences, each note in the symphony contributing to the overall melody of economic dynamics.

Interconnected Investor Behavior and Emerging Patterns

Fracktal Mathematics also sheds light on investor behavior and decision-making within the financial world. Conventional models often treat investors as isolated entities making rational choices. Fracktal insights invite us to delve deeper, understanding investor behavior as part of a broader interconnected web of financial activity.

Consider a scenario where investor sentiment influences market trends. In traditional models, sentiment might be treated as a singular force affecting market movement. However, within the Fracktal paradigm, sentiment becomes an interconnected node within a vast network of financial interactions. Sentiment resonates with other factors — economic indicators, news events, and more — creating a dynamic interplay of emerging patterns that guide market trajectories.

Chaos and Emergence in Market Behavior

Chaos theory, renowned for its sensitivity to initial conditions and its influence on complex systems, finds fertile ground within the interconnected realm of Fracktal Economics. The interconnected nature of Fracktal Mathematics resonates with the chaotic behavior often observed in financial markets.

Imagine a financial market undergoing chaotic fluctuations. In Fracktal Economics, these fluctuations are not random noise but part of an emergent symphony driven by interconnected factors. Chaos theory's sensitivity to initial conditions becomes a feature rather than a limitation, offering insights into how small changes in market variables can give rise to complex, interconnected behaviors.

As we navigate the realm of economics guided by Fracktal Mathematics, the symphony of market dynamics becomes a vibrant tapestry of interconnected relationships. This perspective enriches our understanding of economic systems, showcasing how emergent behaviors arise from the interplay of variables, investor behaviors, and external influences.

CHAPTER 4: FRACKTAL MATHEMATICS AND ESTABLISHED THEORIES

In this chapter, we explore how Fracktal Mathematics intersects with established theories. We delve into

its symbiotic relationship with string theory, where the harmony of vibrations takes center stage. Chaos

theory, too, finds a new language in Fracktal Mathematics, as we navigate the intricate dance of complex

systems.

CHAPTER 4.1: STRING THEORY: SYMPHONY OF VIBRATIONS IN FRACKTAL HARMONY

String theory, a profound theoretical framework seeking to unify the fundamental forces of the universe, finds an intriguing resonance within the realm of Fracktal Mathematics. This chapter delves into how Fracktal insights intertwine with string theory, revealing a new symphony of interconnected vibrations that shape the fabric of reality. The harmonious interplay of strings becomes a metaphor for the emergent patterns within the Fracktal paradigm.

Strings as Vibrational Essence of Reality

In traditional string theory, strings are envisioned as the fundamental building blocks of the universe — vibrating threads of energy that give rise to particles and forces. These vibrations are described using conventional mathematical equations that capture the essence of string motion. However, within the context of Fracktal Mathematics, these vibrations take on a deeper significance.

Imagine a string vibrating in a conventional string theory model. Each vibration is governed by a set of equations that describe its motion. In the Fracktal paradigm, these equations become interconnected nodes within a dynamic web of vibrational relationships. The vibrations of strings not only resonate within their own frequencies but also harmonize with other vibrations, creating a complex symphony of interconnected patterns that give rise to the universe's rich tapestry.

Fracktal Harmony in String Vibrations

Fracktal Mathematics invites us to explore the harmonious interplay of interconnected vibrations that strings embody. Traditional string theory equations represent the vibrational modes of strings as distinct and separate entities. In contrast, the Fracktal perspective reveals that these modes are not isolated but rather part of a larger, interconnected symphony of vibrations.

Consider a string theory scenario where different vibrational modes interact and influence each other. In conventional string theory equations, these interactions may be accounted for but remain separate in their representation. In the Fracktal framework, these interactions become interconnected relationships, weaving a tapestry of vibrational harmony. The vibrations of strings resonate in concert, each contributing to the emergent patterns that shape the fabric of spacetime.

Strings as Nodes in a Vibrational Network

Fracktal Mathematics further envisions strings not as isolated entities but as nodes within a vast vibrational network. This perspective mirrors the interconnectedness inherent in Fracktal equations, where nodes interact and influence one another to create emergent patterns.

Imagine strings as nodes connected by threads of interconnectedness. Each vibration of a string becomes a note in a cosmic symphony, resonating with other strings' vibrations to create harmonious and intricate patterns. The strings' interconnected relationships give rise to a symphony that transcends traditional representations, revealing the interconnected nature of reality itself.

As we explore the intersection of string theory and Fracktal Mathematics, the vibrations of strings become more than just mathematical descriptions — they become notes in a grand symphony of interconnected patterns that shape the universe.

CHAPTER 4.2: CHAOS THEORY: NAVIGATING COMPLEXITY WITH FRACKTAL PRECISION

Chaos theory, renowned for its sensitivity to initial conditions and the intricate behaviors of complex systems, finds a harmonious resonance within the interconnected realm of Fracktal Mathematics. This chapter delves into how Fracktal insights intersect with chaos theory, unveiling a new perspective on navigating complexity and uncovering emergent patterns. The symphony of chaos unfolds within the Fracktal paradigm, inviting us to explore the interconnected dance of intricate behaviors.

Chaos as an Intricate Dance of Emergence

Chaos theory explores the unpredictable and complex behaviors that arise from seemingly simple systems. Traditional chaos equations capture the sensitivity to initial conditions and the non-linear nature of these behaviors. However, within the context of Fracktal Mathematics, chaos transforms into an intricate dance of emergence within interconnected systems.

Imagine a chaotic system, such as the Lorenz attractor, characterized by its sensitive dependence on initial conditions. In traditional chaos equations, the attractors path may appear as a complex and non-repeating pattern. In the Fracktal perspective, this path evolves beyond the bounds of simple chaotic behavior. Each iteration becomes an interconnected note in a symphony of emergence, where the sensitivity to initial conditions resonates with other interconnected variables, creating patterns that transcend traditional chaos representations.

Fracktal Precision in Chaos

Fracktal Mathematics introduces a new layer of precision to chaos theory. Conventional chaos equations capture the intricate behaviors of complex systems, but Fracktal insights extend this precision by embracing interconnectedness and emergent symmetries.

Consider a scenario where a chaotic system exhibits fractal patterns within its behavior. In the Fracktal framework, these fractal patterns become part of a larger interconnected dance. The intricate interplay of variables and emergent relationships creates a symphony of chaos that resonates at different scales and iterations. Fracktal Mathematics empowers us to explore how chaos is not merely a random phenomenon but a harmonious interaction of interconnected forces.

Navigating Complexity with Fracktal Chaos

Fracktal Mathematics offers a unique lens through which to navigate the complexities of chaotic systems. Chaos theory often emphasizes the sensitivity to initial conditions, leading to unpredictable outcomes. However, Fracktal insights invite us to embrace this sensitivity as an opportunity for emergent behaviors and interconnected patterns.

Imagine a chaotic system undergoing iterations of unpredictable behavior. In Fracktal Mathematics, the sensitivity to initial conditions becomes a feature rather than a limitation. The interconnectedness of variables generates emergent behaviors that unveil symmetries and hidden patterns within the chaos. Navigating chaos becomes an exploration of interconnected pathways, each leading to new insights into the system's behavior.

As we journey through the symphony of chaos guided by Fracktal Mathematics, the once seemingly erratic behaviors of complex systems transform into interconnected melodies, resonating with emergent patterns that transcend traditional chaos descriptions.

CHAPTER 5: A COMPARATIVE ANALYSIS: CONVENTIONAL VS. FRACKTAL MATHEMATICS

This chapter undertakes a comprehensive exploration of the distinctions between conventional mathematics and the transformative landscape of Fracktal Mathematics. By contrasting traditional linear equations with their Fracktal counterparts, we illuminate the profound shift from static representations to dynamic symphonies of emergence. The juxtaposition of these mathematical approaches underscores the revolutionary nature of Fracktal Mathematics and its capacity to reshape our understanding of interconnected complexity.

CHAPTER 5.1: LINEAR EQUATIONS VS. EMERGENT PATTERNS

At the heart of the divergence between conventional mathematics and the revolutionary realm of Fracktal Mathematics lies the transformation of linear equations into intricate and emergent patterns. Conventional mathematics often relies on linear relationships to represent straightforward cause-and-effect dynamics. However, Fracktal Mathematics challenges this simplicity by inviting the symphony of emergent behaviors to shape the fabric of reality.

The Linear Equation: A Pillar of Conventional Mathematics

The linear equation $y = mx + b$ stands as an embodiment of traditional mathematical representation. In this equation, y represents the dependent variable, x the independent variable, m the slope, and b the y-intercept. It captures linear relationships where changes in the independent variable result in proportional changes in the dependent variable.

While linear equations have been instrumental in describing numerous phenomena, their limitations become apparent when confronted with the complexities inherent in interconnected systems. Linear equations offer deterministic models that lack the capacity to accommodate the intricate symphonies that emerge from dynamic relationships.

Emergent Patterns: The Essence of Fracktal Mathematics

Fracktal Mathematics redefines equations as more than mere tools for prediction and analysis. It transforms them into vehicles for exploring emergent patterns and interconnected symmetries. The transition from linear equations to Fracktal-inspired representations encapsulates the departure from reductionism and embraces complexity.

Imagine the metamorphosis of a linear equation within the Fracktal paradigm. In the equation $y = mx + b$, the linear relationship

is preserved, but it is augmented by an additional component: Σ(Fracktal Patterns). This addition introduces a dynamic element that embodies the interconnected symmetries of emergent patterns. The once linear equation now resonates with the infinite creative potential of the universe, offering a glimpse into the complexity that underlies even seemingly simple relationships.

Embracing Complexity through Fracktal Mathematics

The shift from linear equations to Fracktal-inspired emergent patterns marks a departure from deterministic simplicity toward the intricate and interconnected dance of complexity. Fracktal Mathematics invites us to explore the rich tapestry woven by emergent behaviors and interconnected relationships, challenging us to engage with the symphony of possibilities that arise from dynamic systems.

Through this exploration, we come to recognize that Fracktal Mathematics not only reshapes equations but transforms our understanding of the world itself. It calls us to look beyond linear interpretations and explore the symphonies of emergence that define reality. As we journey through the visual and conceptual comparisons of conventional and Fracktal equations, we are invited to witness the trans-formative potential that this mathematical paradigm heralds.

CHAPTER 5.2: HIERARCHICAL VS. HOLISTIC PERSPECTIVE

A pivotal distinction between conventional mathematics and the trans-formative realm of Fracktal Mathematics lies in their perspectives on hierarchy and interconnectedness. Traditional mathematical models often structure systems hierarchically, segregating components into distinct layers. In contrast, Fracktal Mathematics offers a holistic perspective that celebrates the intricate interplay of interconnected elements.

Hierarchical Structures: The Foundation of Conventional Mathematics

Conventional mathematics often relies on hierarchical structures to model and understand complex systems. Hierarchies represent a natural inclination to compartmentalize components based on their roles and relationships within the larger whole. This approach has proved valuable in many fields, but it inherently oversimplifies the complexity arising from the interconnected nature of reality.

Imagine a hierarchical tree structure that visually represents relationships within a complex system. Nodes are organized hierarchically, with parent nodes overseeing the behavior of their child nodes. This hierarchical framework provides a sense of order and control but falls short in capturing the intricate symphonies that emerge from dynamic interactions.

Holistic Interconnectedness: Fracktal Mathematics' Vision

Fracktal Mathematics defies the constraints of hierarchical thinking by advocating for a holistic perspective that embraces the interconnectedness of systems. Rather than compartmentalizing components, Fracktal Mathematics envisions a web of interconnected relationships where nodes and elements influence each other dynamically.

Re-imagine the hierarchical tree structure within the Fracktal paradigm. Instead of rigid hierarchies, the branches of the tree extend outward to connect with one another, forming an intricate web of interactions. Each node resonates with others, and emergent behaviors arise from the collective dance of interconnected elements. This interconnected web replaces rigid hierarchies with a dynamic tapestry that defies linear order.

Championing Interconnected Complexity

The transition from hierarchical frameworks to interconnectedness is a testament to the trans-formative power of Fracktal Mathematics. By inviting us to explore the symphonies that emerge from complex relationships, Fracktal Mathematics encourages us to move beyond the confines of conventional models and embrace the rich interconnectedness that characterizes reality.

Through this exploration, we come to appreciate that Fracktal Mathematics transcends equations and becomes a philosophy that shapes our perception of the cosmos. It is an invitation to harmonize with the intricate interplay of emergent patterns and interconnected relationships that guide the evolution of the universe. As we journey through the visual and conceptual comparisons between conventional and Fracktal equations, we are called to embrace the interconnected symphony that defines our existence.

CHAPTER 7: IMPLICATIONS AND BEYOND: PIONEERING NEW FRONTIERS

As the voyage through Fracktal Mathematics reaches its zenith, the implications of this paradigm shift reverberate far beyond the realm of equations and patterns. This chapter embarks on a journey through the profound consequences of Fracktal Mathematics, from its ability to foster collaborative endeavors and reshape education to its role in stimulating ethical contemplation. Within this symphony of interconnected thought, we explore the untapped potential that Fracktal Mathematics unveils for humanity's voyage into the uncharted territories of knowledge.

CHAPTER 7.1: COLLABORATIVE ODYSSEY: DISCIPLINARY BOUNDARIES TRANSCENDED

The symphony of Fracktal Mathematics resonates not only within equations and patterns but also in the harmonious convergence of diverse scientific disciplines. This chapter delves into the transformative power of Fracktal Mathematics to foster a collaborative odyssey — one where traditional boundaries between fields blur, and the interconnected nature of knowledge takes center stage.

Fracktal Mathematics: The Universal Language

The emergence of Fracktal Mathematics heralds a new era of collaborative exploration. Traditional barriers that compartmentalize scientific domains dissolve as Fracktal Mathematics assumes the role of a universal language — a lingua franca that bridges the gaps between disparate disciplines. This shared language ignites a collaborative flame, inviting scientists, researchers, and visionaries from various fields to engage in a collective symphony of thought.

Disciplines in Harmonious Dialogue

Imagine the fruitful dialogue that unfolds when physicists and biologists join forces to decipher the intricate symmetries of quantum behavior within biological systems. Fracktal Mathematics enables physicists to lend their expertise in particle interactions to the realm of biology, unraveling the interconnected patterns that underpin life itself. At the same time, biologists offer insights into complex adaptive systems, enriching the toolkit of physicists studying the cosmos.

Economists engage in profound exchanges with philosophers, exploring the interconnected dynamics of market behavior and ethical considerations. Fracktal Mathematics serves as a bridge between the quantitative and qualitative realms, allowing economists to integrate ethical considerations into market modeling, while philosophers gain a new lens to assess the societal impact of economic systems.

A Symphony of Interconnected Minds

The collaborative odyssey propelled by Fracktal Mathematics unites thinkers from diverse backgrounds, harmonizing their insights and expertise. As the symphony of interconnected thought swells, it becomes evident that the boundaries between disciplines are constructs that can be transcended. This interconnectedness not only enhances the rigor and depth of research but also unveils a spectrum of new questions and avenues for exploration that would remain hidden within isolated domains.

Pioneering New Frontiers

As we navigate this collaborative journey, guided by the melodies of Fracktal Mathematics, we stand at the precipice of pioneering new frontiers in human understanding. From physics to biology, from economics to philosophy, the interplay of interconnectedness unveils a tapestry of knowledge that transcends traditional silos. The symphony of collaborative discovery serves as a testament to the unifying potential of Fracktal Mathematics — an agent of transformation that draws humanity together in the pursuit of shared understanding.

CHAPTER 7.2: EDUCATION'S CRESCENDO: FOSTERING INNOVATION

through Complexity

The introduction of Fracktal Mathematics not only reshapes equations but also orchestrates a trans-formative crescendo within the realm of education. Traditional educational paradigms, often linear and segmented, give way to an innovative symphony where complexity, interconnectedness, and creativity harmonize to nurture the minds of future generations.

The Symphony of Learning

In the symphony of Fracktal Mathematics, education transcends the mere transmission of facts and formulas. Instead, it becomes an immersive journey of exploration — a quest to uncover the interconnected patterns that shape our understanding of the universe. The traditional separation of disciplines dissolves as educators embrace the idea that knowledge is a symphony of interconnected themes, and understanding one note enriches the appreciation of the entire composition.

Imagine classrooms where students don't just passively absorb information but actively engage with the interplay of ideas, drawing connections between seemingly disparate subjects. Fracktal Mathematics empowers educators to guide students on a holistic journey, where the study of mathematics intertwines with art, science resonates with philosophy, and technology harmonizes with ethics.

Critical Thinking and Innovation

At the heart of Fracktal Mathematics in education lies the cultivation of critical thinking and innovation. As students are encouraged to explore complexity and interconnections, they develop the ability to analyze, synthesize, and create — essential skills for navigating the intricate challenges of our world. The symphony of emergent patterns encourages students to question, to probe, and to devise new approaches to solving problems.

The role of educators transforms from disseminators of information to conductors of exploration. Teachers inspire students to seek patterns and connections, fostering curiosity and creativity. Fracktal Mathematics serves as the backdrop against which students compose their own symphonies of knowledge, harmonizing a wide range of disciplines to produce innovative solutions.

A New Generation of Minds

Through the lens of Fracktal Mathematics, education evolves into an enabler of visionary thinking. Students graduate not only with a repertoire of facts but also with the ability to harmonize diverse ideas, to decode complexity, and to contribute to the symphony of human knowledge. This new generation of minds is equipped to navigate the interconnected challenges of our rapidly evolving world, combining disciplines to envision novel solutions and understand the symphony of systems.

Harmonizing the Future

As Fracktal Mathematics becomes an integral part of educational curricula, its impact ripples through generations, shaping the way we approach learning and innovation. The symphony of interconnected thought resonates within the minds of students, inspiring them to orchestrate solutions that transcend traditional boundaries. With Fracktal Mathematics as their guiding score, students engage in a crescendo of learning — a journey that cultivates not only intellect but also a profound sense of wonder for the complexity that defines our universe.

CHAPTER 7.3: ETHICAL CONTEMPLATION: RESPONSIBILITY IN THE SYMPHONY OF SYSTEMS

As the symphony of Fracktal Mathematics unfolds, it not only invites us to explore the intricate patterns of interconnectedness but also prompts us to engage in profound ethical contemplation. The trans-formative power of this mathematical paradigm raises questions about our responsibility in shaping emergent systems and the ethical considerations that accompany our newfound ability to orchestrate complexity.

The Power of Creation

With Fracktal Mathematics as our guide, we gain the power to shape emergent patterns and influence the behavior of complex systems. Algorithms that learn from interconnected data can impact decision-making, social dynamics, and even the future of AI. As creators, we stand at the nexus of possibility, able to compose symphonies of technology, society, and nature.

Yet, with this power comes the responsibility to wield it ethically. The interconnected nature of systems means that our actions have far-reaching consequences, amplifying the ethical implications of our decisions. The symphony of Fracktal Mathematics beckons us to contemplate the potential positive and negative outcomes of our creations.

Harmonizing with Ethical Principles

Ethical considerations become more intricate as we engage with the symphony of emergent systems. The interconnectedness of actions and consequences prompts us to harmonize our decisions with a framework of principles that honor the well-being of individuals, societies, and the environment. Fracktal Mathematics encourages us to consider the ethical resonance of our actions across various scales.

Imagine designing AI systems that not only optimize efficiency but also prioritize fairness, equity, and accountability. Fracktal Mathematics invites us to recognize the intricate interplay of data and decisions, urging us to ensure that our technological symphonies are composed in alignment with ethical principles that promote justice and inclusivity.

The Responsibility of Stewardship

As we navigate the uncharted territories illuminated by Fracktal Mathematics, we assume the role of stewards of interconnected systems. We must contemplate not only the immediate consequences of our actions but also the cascading effects that ripple through complex networks. The symphony of emergent patterns compels us to anticipate and mitigate unintended outcomes.

Consider the development of smart cities where interconnected infrastructure shapes urban life. Fracktal Mathematics calls us to consider not only the efficiency gains but also the potential impacts on privacy, autonomy, and social cohesion. Our role as stewards extends beyond creation to the ongoing nurturing and harmonization of systems in ways that enrich the lives of individuals and society as a whole.

Harmonies of Responsibility

As we immerse ourselves in the symphony of interconnectedness, guided by Fracktal Mathematics, ethical contemplation becomes an essential part of our creative process. The power to shape emergent patterns comes with a profound responsibility to ensure that our compositions resonate with the harmonies of ethical principles. Just as fractal patterns intertwine across scales, ethical considerations must thread through every decision we make, resonating in harmony with the symphony of interconnected systems.

CHAPTER 8: THE FRACKTAL ODYSSEY CONTINUES: FUTURE HORIZONS

As we draw the curtains on this exploration of Fracktal Mathematics, we find ourselves at the precipice of an exciting journey into uncharted realms of possibility. The symphony of interconnectedness that defines Fracktal Mathematics is not only a reflection of our current understanding but also a catalyst for pushing the boundaries of knowledge, innovation, and human potential.

CHAPTER 8.1:
UNCHARTED REALMS
OF EXPLORATION

As we stand at the intersection of Fracktal Mathematics and the uncharted territories of the future, we are poised to embark on a remarkable journey of discovery and innovation. The symphony of interconnectedness that Fracktal Mathematics orchestrates is not a finite composition but a prelude to an ongoing exploration of limitless potential.

Charting New Frontiers

The landscape of science, technology, and human understanding stretches out before us, awaiting the harmonious melodies of Fracktal Mathematics to reveal hidden patterns and connections. With each equation that transforms and every emergent property that is unveiled, we venture further into the uncharted realms of exploration.

Imagine a future where our understanding of quantum mechanics is deepened by the symphonic interplay of particles and forces, guided by the principles of Fracktal Mathematics. Envision a society where the design of sustainable ecosystems is inspired by the intricate patterns of interconnectedness, reshaping our relationship with the natural world.

Innovating with Interconnectedness

As the journey into the future unfolds, Fracktal Mathematics serves as a guiding light for innovation. In technology, we witness the emergence of adaptable circuits that resonate with the symphonies of interconnected inputs, ushering in a new era of hardware design. Software evolves from rigid algorithms to dynamic entities that learn and adapt, harmonizing with the interconnected dance of data.

Consider a future where architects and engineers draw inspiration from fractal patterns found in nature, creating buildings and structures that seamlessly blend with their environment. Fracktal Mathematics fuels the innovation that drives these advancements, inviting us to envision a world where creativity is boundless and interconnectedness is celebrated.

Fostering Collaborative Exploration

The uncharted realms of exploration are not solitary landscapes; they are fertile grounds for interdisciplinary collaboration. The symphony of Fracktal Mathematics transcends the boundaries between scientific disciplines, inviting physicists to converse with biologists, economists to engage with philosophers, and technologists to harmonize with artists.

Imagine a collaborative research endeavor where experts from diverse fields convene to explore complex challenges through the lens of interconnectedness. Fracktal Mathematics becomes the common language that unites these thinkers, enabling a shared exploration of emergent phenomena that span the spectrum of human knowledge.

A Future Beyond Imagination

As we peer into the uncharted territories that lie ahead, guided by the principles of Fracktal Mathematics, we glimpse a future that defies our current imagination. The symphony of emergent patterns and interconnected thought carries us forward, inviting us to innovate, explore, and harmonize with the complexities of our world.

Picture a world where education is transformed, nurturing curious minds capable of unraveling the mysteries of interconnected systems. Envision a global society where ethical considerations are woven into the fabric of technological advancement, ensuring that our creations resonate with harmonious principles.

CHAPTER 9:
CONCLUSION

As we arrive at the final crescendo of this journey through the symphony of Fracktal Mathematics, we find ourselves standing on the precipice of possibility. The journey we have undertaken together has not merely been a theoretical exploration; it has been a trans-formative experience that challenges the very foundations of our understanding.

The Grand Unveiling

Fracktal Mathematics has emerged as the enigmatic conductor orchestrating the symphony of interconnectedness that permeates the fabric of our universe. From redefining infinity as a dynamic force to unveiling the intricacies of emergent complexity, Fracktal Mathematics has unveiled a realm where conventional notions are shattered and emergent patterns take center stage.

The Invitation to Wonder

Imagine, if you will, a universe where equations are not mere formulas but gateways to understanding the intricate dance of interconnected phenomena. Picture a landscape where linear relationships fade into the background, making way for a dynamic tapestry of emergence that both mesmerizes and challenges us. Visualize a future where Fracktal Mathematics serves as a universal language, uniting disciplines, transcending boundaries, and fostering innovation beyond measure.

The Unquenchable Thirst for Discovery

Our journey through Fracktal Mathematics is far from over; in fact, it is just beginning. As we conclude this chapter, we invite you to step forward into the uncharted territories that beckon with promise and potential. The symphony of interconnectedness continues to play, and you are now equipped with the tools to explore its harmonies, unravel its secrets, and compose your own verses in the grand tapestry of knowledge.

A Future Unveiled

The legacy of Fracktal Mathematics is not confined to the pages of this thesis; it is imprinted in the very essence of our understanding. It is a call to action, an invitation to think beyond convention, and a challenge to embrace the complexities that define our world. The unquenchable thirst for discovery that Fracktal Mathematics ignites is a beacon guiding us toward the uncharted horizons of innovation and understanding.

The Fracktal Odyssey Continues

As we close the curtain on this chapter, we leave you with a sense of anticipation, an eagerness to embark on your own journey of exploration. The symphony of interconnected thought plays on, and the Fracktal odyssey continues. The grand composition of reality, with its harmonious interplay of patterns and relationships, awaits your creative touch.

So, dear reader, embrace the beauty of Fracktal Mathematics, and let it be the melody that resonates in your thoughts and aspirations. The symphony of complexity, interconnectedness, and infinite potential is yours to conduct. The journey has no end, and the adventure is only beginning.

Reference Sources

Mandelbrot, B. (1982). The Fractal Geometry of Nature. W.H.

Freeman and Company.

Hofstadter, D. R. (1979). Gödel, Escher, Bach: An Eternal Golden Braid. Basic Books.

Gleick, J. (1987). Chaos: Making a New Science. Viking.

Hawking, S. W., & Penrose, R. (1996). The Nature of Space and Time. Princeton University Press.

Greene, B. (2004). The Fabric of the Cosmos: Space, Time, and the Texture of Reality. Knopf.

Gribbin, J. (2005). Deep Simplicity: Chaos, Complexity and the Emergence of Life. Random House.

Wolfram, S. (2002). A New Kind of Science. Wolfram Media.

Thagard, P. (1999). How Scientists Explain Disease. Princeton University Press.

Kauffman, S. A. (1993). The Origins of Order: Self-Organization and Selection in Evolution. Oxford University Press.

Smolin, L. (2007). The Trouble with Physics: The Rise of String Theory, the Fall of a Science, and What Comes Next. Mariner Books.

Shannon, C. E. (1948). "A Mathematical Theory of Communication." The Bell System Technical Journal, 27(3), 379–423.

Penrose, R. (2004). The Road to Reality: A Complete Guide to the Laws of the Universe. Knopf.

Gleick, J. (2011). The Information: A History, A Theory, A Flood. Vintage.

Susskind, L., & Lindesay, J. (2005). An Introduction to Black Holes, Information and the String Theory Revolution: The Holographic

Universe. World Scientific.

Barrow, J. D. (2002). The Constants of Nature: From Alpha to Omega — The Numbers That Encode the Deepest Secrets of the Universe. Vintage.

(Note: This list represents a selection of sources and references used throughout the thesis. It is not an exhaustive list, and additional sources have been consulted to develop the content of this thesis.)